10|98

FARADA

D0362217

In the same series by the same authors:

Curie in 90 minutes
Darwin in 90 minutes
Einstein in 90 minutes
Galileo in 90 minutes
Halley in 90 minutes
Mendel in 90 minutes
Newton in 90 minutes

John and Mary Gribbin

FARADAY
(1791–1867)
in 90 minutes

Constable · London

First published in Great Britain 1997
by Constable and Company Limited
3 The Lanchesters, 162 Fulham Palace Road
London W6 9ER
Copyright © John and Mary Gribbin 1997
The right of John and Mary Gribbin to be identified
as authors of this work has been asserted by them
in accordance with the Copyright,
Designs and Patents Act 1988
ISBN 0 09 477100 6
Set in Linotype Sabon by
Rowland Phototypesetting Ltd,
Bury St Edmunds, Suffolk
Printed in Great Britain by
St Edmundsbury Press Ltd,
Bury St Edmunds, Suffolk

A CIP catalogue record for this book
is available from the British Library

921
FARADAY
1997

Contents

ACKNOWLEDGEMENT

Thanks to Bill Murray for help in tracking down reference material.

Faraday in context

Michael Faraday was one of the first modern scientists. He discovered the principles of, and invented, both the electric motor and the generator, without which modern society could not function. In order to explain his discoveries and inventions he came up with the idea of a field of force, a concept which now underpins our understanding of everything from the gravitational force holding the Universe together to the forces operating between quarks inside the protons and neutrons that make up the nuclei of atoms of ordinary matter. Yet when Faraday first became interested in chemistry, early in the nineteenth century, even the concept of atoms was a controversial idea, and chemistry had scarcely emerged from its origins in alchemy.

Only in the final quarter of the eighteenth century, thanks to the work of Antoine Lavoisier in France and Joseph Priestley in England, had it become clear that the process of burning involves something from the air (oxygen) combining with the thing that is burning – not a loss of 'phlogiston' from the thing being burnt. In later experiments Lavoisier began to show how compounds like carbon dioxide and water are produced by combustion, and also during the respiration of living things – an early indication that life does not involve any unique 'life force'. His career was cut short (literally) by the guillotine during the French Revolution.

In 1805 Lavoisier's widow married an American-born scientist, Benjamin Thomson, who had been made Count Rumford in 1791 for his services to the Elector of Bavaria. Rumford's wide-ranging career (he had had to leave America after fighting on the losing side in the War of Independence) had made him rich and influential, and during a period spent in England at the end of the eighteenth century and the beginning of the nineteenth, he had been instrumental in establishing London's Royal Institution (usually known as the RI), in 1800. This would be where Faraday carried out all his work; it was Rumford who would be responsible for appointing Humphry Davy as the RI's first director and, as we shall see, it was Davy who would give Faraday his first big break.

The discovery that an electric current could be produced without using animal tissue (another indication that there is no unique life force) was announced by the Italian Alessandro Volta at the end of the eighteenth century. Before then, people had known about what is now called 'static' electricity, which can be produced by friction. This is most familiar today as the electric charge that builds up in your body if you walk around on a carpet made of synthetic fibres on a dry day; it can be enough to give you a sharp shock when you touch a metal object and the electricity is discharged. This is the kind of electricity that was studied by Benjamin Franklin, who lived from 1706 to 1790, and was lucky not to have been killed halfway through that long span in one of his famous kite-flying experiments during thunderstorms.

Interest in the kind of electricity that involves a current – a flow of charge – was fired in the eighteenth century by studies of electric fish and eels. Luigi Galvani, a compatriot and friend of Volta, found that the legs of a dissected frog would twitch when given a jolt of static electricity, and that the same sort of twitching would occur when the leg was laid across two different metals (such as copper and iron). He concluded (wrongly, it turned out) that electricity was being made in the muscles of the legs by a process like that by which an eel makes electricity. It was Volta who showed that the production of an electric current in this way does not require living (or formerly living) tissue at all, but will happen when two metals are in a solution of salt – the frog's legs were only acting as a damp, salty connection between the iron and the copper.

Volta was working at the University of Pavia in Lombardy at the end of the eighteenth century when he made this discovery. He found that a steady electric current could be produced by making a pile of alternating copper and iron (or zinc) discs, separated by layers of cloth soaked in salt solution. When the top and bottom of the pile were joined with a wire, a current would flow.

Because of the Napoleonic Wars, the state of Lombardy passed repeatedly between French and Austrian hands at this time, and in 1799 the Austrians gained control and closed the University of Pavia. So Volta sent news of his discovery to London, where it was published, in 1800, by the Royal Society. The same year, the French retook Lombardy and the University was reopened; Napoleon invited Volta to Paris to demonstrate his pile, and made him a count in recognition of the work. Volta's discovery was sensational, and (among other things) inspired a burst of work by Davy, once he was securely installed at the RI.

Davy pioneered the development of electrochemistry, using the current provided by a 'voltaic pile' (what we would now call a battery) to break compounds into their constituent elements. Using this technique, he discovered several 'new' elements, including potassium and sodium. In 1808, Davy received a prize of 3000 francs offered by Napoleon for the best research on current electricity (the fact that England and France were at war having little effect on the flow of scientific ideas and honours). The kind of battery that Faraday would use in much of his early work was developed by William

Cruikshank, and consisted of copper and zinc plates soldered together in pairs, sealed into grooves in a wooden trough using wax, and immersed in acid – not that different from a modern car battery.

So the first decade of the nineteenth century was an exciting time for anyone interested in science, with news of new elements being discovered, the invention of the electric battery, and speculation about what this mysterious thing called electricity was. In London, interest in science in fashionable society was fired by Davy's lectures at the RI. This was somewhat ironic, because Rumford's aim in establishing the RI was to provide a place where ordinary people could learn about science. The snag was, ordinary folk had no money, and the RI was kept afloat by the contributions of fashionable young women, in particular, attracted to the lectures not so much by love of science as by Davy's good looks and animal magnetism. Nevertheless, Rumford's hopes for the RI were soon to be realized in the most dramatic fashion.

Life and work

Faraday was born on 22 September 1791. He had an elder brother, Robert, born in 1788, and an elder sister, Elizabeth, born in 1787. The family came from what was then Westmorland, in the north of England. James Faraday brought his wife, Margaret, and their first two children south in search of work in 1791. He was a blacksmith, and by all accounts a good one, but suffered from bad health and barely earned enough to keep the family together. They lived briefly in Newington (then a village in Surrey, now part of the urban sprawl of south London), where Michael was born, then moved north of the River Thames to cramped rooms over a coach house in Jacob's Well Mews, near Manchester Square. There, another child, Margaret, was born in 1802.

Although desperately poor, the family was a loving and happy one, at least partly thanks to their religious faith. The Faradays were members of a sect known as the Sandemanians, which originated in the 1730s in a breakaway from the Scottish Presbyterians. (The

Sandeman who gave his name to the sect, incidentally, was a relation of the Sandemans who gave us the port.) Sandemanians regarded themselves as members of the only true Church, and therefore assured of salvation, a belief which made it easier for them to tolerate the hardships of the present world. They tended not to socialize much outside their own sect (although they were by no means as exclusive as, say, the Plymouth Brethren), were not interested in worldly goods and wealth, and made unostentatious donations to charity.

All this provided Faraday with a strict moral code, and a certain serenity with which to cope with life's vicissitudes. It also encouraged him in his scientific work, through the belief that an understanding of God's 'book of nature' was as important as an understanding of the Bible. The Sandemanians were not evangelical, believing that those who belonged in their community would naturally find a way to them; hardly surprisingly, the sect never had more than a few hundred adherents and has now essentially died out. We shall not say much more about Faraday's

Sandemanian faith in this book, but it is worth remembering as an important part of the background to the story.

Faraday received only the most basic education, in the traditional 'three Rs' of reading, (w)riting and (a)rithmetic. 'My hours out of school,' he later wrote, 'were passed at home and in the streets.' In 1804, when he was 13, it was time for Faraday to start contributing to the family income. He ran errands for George Riebau, a bookseller and bookbinder who had a shop in Blandford Street, close to where the Faradays lived, and just off Baker Street, mythical home of Sherlock Holmes.

One of Faraday's main duties was to deliver newspapers and fetch them back to the shop – at that time, some people could not afford to buy newspapers, but paid a smaller sum for the loan of one, which had to be returned after being read. Riebau was a kindly man, and through this work Faraday was introduced to the world of books. A year later, when he was 14 and had to learn a trade, it was natural for him to be apprenticed to Riebau to learn bookbinding.

Little is known about Faraday's life during

his first four years with Riebau, but from the many volumes still in existence which he bound for himself later in life, it is clear that he learned his trade well. The fact that he became a superb experimental scientist who built his own apparatus and carried out delicate experiments with skill must owe something to the manual dexterity that he developed for bookbinding.

But it wasn't all work. Some idea of the atmosphere that prevailed at George Riebau's premises may be gleaned from the fact that Faraday had two fellow-apprentices there, one of whom went on to become a professional singer, while the other achieved success on the boards as a comedian. They worked hard, but they also played hard, and there were opportunities for each of the young men to develop their own interests in what was, in effect, a happy family unit. Faraday soon moved in to live on the premises.

In Faraday's case, 'play' largely amounted to study. He read voraciously from the stock in Riebau's shop and the books that came in for binding, and he would carefully copy out interesting passages from the volumes that passed through his hands. When he became interested in chemistry, he bought a standard four-volume introduction to the subject, which he dismantled and rebound with blank pages between the pages of text, so that he could make his own notes as he slowly came to grips with the topic.

His fascination with electricity was fired by reading the article on the subject in a copy of the third edition of the *Encyclopaedia Britannica* that was brought in for binding. As well as reading, Faraday carried out experiments, building apparatus (including a voltaic pile) out of any bits and pieces he could lay his hands on. But these electrical experiments did not take place until 1812, by which time Faraday was well and truly hooked on science.

The turning-point came at the beginning of 1810, when Faraday, now 18, saw an advertisement for a series of evening lectures on natural philosophy – the term then used for what we now call science. The lectures were open to members of the grandly styled City Philosophical Society, actually a fledgling organization founded two years before by a group of young men eager for self-improvement. Membership cost a shilling; Faraday's subscription was paid by his brother Robert (by now working as a black-smith). Over the next two years he attended lectures on a variety of scientific topics, taking careful notes which he wrote up in elaborate detail back in Blandford Street.

At the Society he made new friends (notably Benjamin Abbott, with whom he carried on an extensive correspondence, much of which survives), and gained in confidence sufficiently first to join in discussions at the meetings and then to give a lecture (on electricity) himself. But he was also painfully aware of his own lack of education and of his rough ways. The correspondence with Abbott was deliberately started as an exercise in improving Faraday's skill at written communication, and he persuaded another new friend, Edward Magrath, to spend two hours a week helping him improve his grammar, spelling and punctuation. This tuition was to continue for seven years.

Faraday's father did not live to see much of this development in his youngest son; he died in October 1810, when he was 49 and Michael was just 19 (Faraday's mother, Margaret, survived until 1838, dying at the age of 74). Faraday himself cannot even have dreamed where his interest in science would take him. But the next great step came in 1812, when he was a few months short of his 21st birthday, and increasingly aware that his time as an apprentice would soon be at an end; and it was as a direct result of his involvement with the City Philosophical Society.

This is where stories of Faraday the bookbinder's apprentice usually begin. By now, Faraday had put together four bound volumes of his notes from meetings of the Society. Riebau, proud of the presence of such an enthusiastic natural philosopher in his household, used to show these volumes to his friends and customers. One of those customers, Riebau later recalled, was a young Mr Dance, who thought the work so remarkable that he asked if he could borrow the books so he could show them to his father. The upshot was that the elder Mr Dance sent tickets for Faraday to attend lectures given by

ALBUQUERQUE ACADEMY LIBRARY

Humphry Davy at the Royal Institution. As was his habit, the enthralled Faraday took careful notes, which he wrote up, with accompanying drawings of the experiments carried out by Davy, and bound. Riebau tells us that 'this he took also to the Above Gent. [Mr Dance senior] who was well pleased.'

In the spring of 1812, Faraday attended four lectures at the RI. It was almost too much of a good thing. The experience reinforced his interest in science, and he now desperately wanted to make a career out of it, but there seemed to be no prospect at all of this happening. Faraday's apprenticeship ended on 7 October 1812, a couple of weeks after his 21st birthday, and he began working as a bookbinder for a Mr De La Roche. De La Roche seems to have been a difficult master, but it is hard to imagine anyone living up to George Riebau – and, to be fair, Faraday's heart was certainly not in his work. He wrote to a friend that he was 'now working at my old trade, the which I wish to leave at the first convenient opportunity'. He even wrote to Sir Joseph Banks, the President of the Royal Society, asking if there was any way to get even the most menial job in science; Banks didn't bother to reply.

Before the end of October, though, Faraday had a stroke of luck. Davy was temporarily blinded by an explosion, and needed somebody who could act as his secretary for a few days – preferably someone who knew a bit about chemistry. Almost certainly on the recommendation of the elder Mr Dance, Faraday got the job. How he got time off from his bookbinding duties we do not know; it can hardly have improved the relationship with his new employer. But Davy soon recovered his sight, Faraday went back to work, and the future closed in on him again. By December, he was trying to build on the contact with Davy, writing to him and sending him the bound volume of notes from Davy's own lectures earlier that year, again begging to be considered for even the most menial of scientific posts.

Davy was impressed, both by the bound lecture notes and by Faraday's enthusiasm, but there were simply no openings at the RI, which was being run on a shoestring budget. In February 1813, though, came the final piece of good fortune that turned a bookbinder's apprentice into one of the greatest scientists of all time. For ten years, a certain William Payne had served as laboratory assistant at the RI, without achieving any distinction in that role. Never a particularly temperate character, Payne now got involved in a public brawl, and was dismissed. Davy sent for Faraday and offered him the job, at a guinea a week, with accommodation provided in two rooms at the top of the RI building in Albemarle Street, candles and fuel included. With enough presence of mind to insist that the RI should provide him with aprons and agree that he could use the apparatus there for his own experiments in his spare time, Faraday accepted. He was officially appointed on 1 March 'to fill the situation lately occupied by Mr Payne on the same terms'.

From the outset, Faraday's duties as Davy's assistant at the RI involved rather more than mere bottle-washing. He was given routine tasks to carry out, including the extraction of sugar from beet (in the nineteenth century, the development of a home-grown source of sugar was of great economic importance to Britain). He worked with Davy on the violently reactive nitrogen chloride, and wrote to Abbott in tones of glee about the explosions that resulted. Because of his skill at making apparatus, and making it work, he was soon in demand as the demonstrator at the lectures that were the RI's *raison d'être*. He continued, whenever possible, to attend the meetings of the City Philosophical Society. But just as Faraday was settling in to a happy routine at the RI, his life was changed again.

Davy, to the despair of many of the fashionable ladies of London, had married a wealthy widow in 1812 (the same year that he was knighted by the Prince Regent). He had a fancy to take his bride on a tour of Europe, undaunted by the fact that Britain and France were at war – after all, he had been awarded the Napoleon Prize for his work. Scientific exchanges between the two countries were, if not exactly routine, at least possible in those days, and when Davy applied for a special passport to enable him to travel on the continent for scientific reasons (one of the aims of the expedition was to study the chemistry of volcanic lavas), the French agreed to his request. Davy invited Faraday to go along with them. It would mean resigning his post at the RI, but with a guarantee that he would get it back when they returned to England. Faraday, who had never travelled more than 12 miles from the centre of London, leaped at the opportunity.

The party set off on 13 October 1813. It consisted of Sir Humphry and Lady Jane Davy, Lady Jane's maid and Faraday. Davy's valet was supposed to accompany them, but, perhaps deterred by the prospect of Napoleonic France, refused at the last minute. Davy asked Faraday if he would mind taking on some of the valet's duties, just until they got to Paris, where a replacement would be hired; again, he agreed. This was to lead to considerable friction, because somehow a suitable replacement valet was never found, and Faraday was in the uncomfortable position of being both colleague and servant to Davy throughout the eighteen-month trip. It wasn't so bad with Sir Humphry, but Lady Jane was a class-conscious woman who believed in keeping servants firmly in their place. But for

Faraday the delights of the trip amply outweighed these inconveniences.

Faraday kept a journal in which those delights were recorded, starting with his awe-struck impressions of the 'mountainous nature' of the Devonshire countryside through which they travelled down to Plymouth to join their ship. As eager as ever for self-improvement, Faraday made the most of the European trip. He met leading scientists in France, Switzerland and Italy; he saw Paris and Rome, real mountains, and a waterspout in the Mediterranean. In Paris, he worked with Davy and French chemists on the identification of a newly discovered element, iodine. In Florence, he saw the telescope that Galileo had used to discover the moons of Jupiter.

During these travels, the wars continued. Napoleon was defeated at Leipzig and forced to abdicate in 1814, but escaped from Elba and returned to France in February 1815. Perhaps even Davy's insouciance was shaken by this, since his party returned home in some haste in April that year, travelling back from Italy via Switzerland and Germany, avoiding France. Napoleon's ultimate defeat took place a few weeks later, on 18 June 1815, at Waterloo.

But all of this seems to have passed Faraday by. He never took any interest in politics, and the matters of the world meant little to him, happy as he was with his science, his religious faith and (soon) his wife. Even so, the young man who returned to England in 1815 was far more sophisticated than the laboratory assistant who had left in the autumn of 1813. He had learned to read French and Italian, and to speak French adequately; he had carried out real scientific research, as the partner (albeit the junior partner) to Davy, not merely as his assistant. He was now a real natural philosopher, and his status was recognized by the RI, where instead of getting back his old job as promised, he was made Superintendent of the Apparatus and Assistant in the Laboratory and Mineralogical Collection, with remuneration of 30 shillings per week (an increase of almost 50 per cent on his previous income), and better rooms at the RI as well.

Davy, seduced by the delights of society, spent less time at the RI than before, and was often out of London. Faraday spent more time than ever helping the lecturers, and he worked particularly closely with the Professor of Chemistry at the RI, William Brande – both in his lecturing and in the analytical work that he carried out for commercial clients. Faraday began to carry out research on his own, and published a series of modest papers. Over a three-year period from January 1816, he gave a total of sixteen lectures to the City Philosophical Society covering the entire subject of inorganic chemistry.

All the while, he continued to read and learn as part of his programme of self-improvement. There was as yet no hint, though, of the way Faraday would revolutionize scientific thinking. He had developed into a steady, reliable chemist, not considered to be in the same league as Davy, but sound enough. He had settled down, and in 1821, on 12 June, he married Sarah Barnard, another member of the Sandemanian community (five years later, Michael's sister Margaret married Sarah's brother John). The first hint of what was to be Faraday's masterwork came in the year that he married – but it proved a mixed blessing, and something of a false dawn.

In 1820, the Danish scientist Hans Christian Oersted had discovered the first link between electricity and magnetism. He had found that when a magnetic compass needle was held over a wire carrying an electric current, the needle was deflected to point across the wire, at right angles. If the needle was held under the wire, it pointed in the opposite direction, but still at right angles to the current flowing in the wire. This was a sensational discovery, because it seemed to imply that the magnetic force associated with the current acted in a circle around the wire. It did not push or pull the tiny magnet of the compass towards or away from the wire, and seemed to be completely different from the push–pull effects of static electricity and magnetism, and the pulling of gravity.

Scientists across Europe attempted to explain the phenomenon, and began to carry out experiments in electromagnetism. In Britain, William Wollaston developed a theory that the electric current in a wire must travel in a helix down the wire, like a child sliding down a helter-skelter, and that this circulating current was responsible for the circular magnetic force. He reasoned that if that were so, a wire carrying a current ought to turn on its axis, like a spinning top, if it were brought near a magnet. In April 1821, he went along to the RI and carried out some experiments with Davy to search for this effect, but without success. Faraday was not present at the experiments, but joined in the discussion of their significance afterwards.

Faraday's own interest in electromagnetism was triggered when, a little later in the year, he was asked to write a historical account of the phenomenon for the journal *Annals of Philosophy*. In order to do a thorough job, he repeated all the experiments described by the scientists whose work he was summarizing. He realized that there really was a circular force associated with an electric current, and came up with an ingenious demonstration of the effect. In one experiment, a wire carrying a current was made to circle round and round a fixed magnet, while in a variation on the theme the wire carrying the current was fixed and the magnet moved round the wire. 'The effort of the wire,' he wrote, 'is always to pass off at a right angle from the pole [of the magnet], indeed to go in a circle round it.'

This was completely different from the effect predicted by Wollaston, and which Wollaston and Davy had failed to find. But when Faraday's paper announcing the discovery (his first really important paper) was published in October 1821 (just after his 30th birthday), many people who had not followed the story closely but vaguely remembered Wollaston talking about rotations leaped to the conclusion that at best Faraday had simply developed Wollaston's idea, and at worst he might be guilty of stealing Wollaston's work. Even Davy, who ought to have known better, felt that Faraday had behaved badly, and this was the beginning of a rift between the two men that was never healed. It may be that Davy resented the prospect of being overshadowed by his former lab assistant and part-time valet; whatever his motives, when Faraday was proposed as a Fellow of the Royal Society in 1823, Davy led the opposition to his election. Davy was then President of the Royal Society, but in spite of his opposition Faraday was elected a Fellow in January 1824; there was a single black ball among the votes.

As a result of all this unpleasantness, Faraday cut himself off even more from society, and became almost a recluse. He and his wife already lived literally above his place of work, and life revolved around the domestic world upstairs, the laboratories and lecture room below, and the Sandemanian church. Not that Faraday didn't enjoy life, either upstairs or downstairs. Family visitors, especially his nieces (Michael and Sarah had no children of their own) later recalled the fun they had on visits, and how Uncle Michael would entertain them with chemical tricks in the laboratory – which was also a convenient place to brew the ginger wine for Christmas.

Faraday's demonstration of how a wire carrying a current could be made to rotate in a magnetic field used simple apparatus that was soon copied, and the effect was studied all over Europe. It made his name, as his prompt election to the Royal Society shows, and it led directly to the development of the electric motor. Sixty years after the demonstration of Faraday's table-top experiment, electric trains were running in Germany, Britain and the United States. If he did nothing else, Faraday would now have been assured of a place in scientific history.

In fact, as far as electromagnetism was concerned, he did very little else for the next ten years. He returned to chemistry, and became the first person to liquefy chlorine in 1823, using a technique devised by Davy. This was quite an achievement at the time, and Faraday went on to liquefy other gases that had previously stubbornly resisted all attempts to make them do so. But this was hardly in the same league as his work on electromagnetism. In 1825 he discovered the compound now called benzene, which is now known to have a ring-shaped structure of linked carbon atoms that is very important in molecules of life, such as DNA; but this was no more than a curiosity in his lifetime. Throughout much of the 1820s, Faraday was involved in a mammoth project to find ways to manufacture better kinds of glass for use in navigational instruments.

By 1825, Faraday's prestige was so high,

and his value to the RI so clear, that it seemed appropriate to give him a new status at the Institute. Davy had originally held the joint posts of Professor of Chemistry and Director of the Laboratory at the RI. He had already given up the former in favour of Brande, and now he retired as Director of the Laboratory. (Although he was not yet 50, Davy seems to have lost interest in science; he became ill in 1827, and died of a heart attack in 1829.) At Davy's suggestion (and in spite of the damage done to their friendship by the Wollaston incident), on 7 February 1825 Faraday became Director of the Laboratory, although there were initially no funds to give him an increased remuneration. He had already made his debut as a lecturer at the RI the previous year, standing in for Brande on a course of chemistry for medical students, and now he developed his interest in lecturing activities in two directions.

First, he introduced a series of Friday evening meetings, which grew into an institution in their own right, the Friday Evening Discourses, at which different speakers, but often Faraday himself in the early years, reported the latest developments in science to a general audience of paying customers. The tradition continues to this day. He then started a series of lectures for children, to be given each Christmas. Again, the tradition continues to this day, and now the Christmas Lectures, given by a different guest lecturer each year, are broadcast on TV. Together, the Friday Evening Discourses and the Christmas Lectures have introduced generations of people to the wonders of science. Along the way, Faraday became a superb and famous lecturer who could always draw a crowd. Between 1825 and 1862, when he retired, Faraday gave more than a hundred of the Friday lectures himself.

All this activity, including the lengthy project on glass, brought money in to the RI and saved it from extinction in the late 1820s. Without Faraday, the Institution would probably have gone bust, and this largely explains why he expended so much effort on the commercial side of the RI's activities instead of concentrating on electromagnetism. Faraday's feelings for the RI were clearly expressed in his response to an offer from the

University of London of the Chair of Chemistry, in 1827. Although flattered, he replied that:

> I think it a matter of duty and gratitude on my part to do what I can for the good of the Royal Institution in the present attempt to establish it firmly. The Institution has been a source of knowledge and pleasure to me for the last fourteen years, and though it does not pay me in salary for what I *now* strive to do for it, yet I possess the kind feelings and good-will of its authorities and members, and all the privileges it can grant or I require; and, moreover, I remember the protection it has afforded me during the past years of my scientific life . . . I have already (and to a great extent for the sake of the Institution) pledged myself to a very laborious and expensive series of experiments on glass . . .

And besides, Faraday loved the opportunity the RI gave for public lecturing. His loyalty was rewarded in 1833, when a new endowment to the RI made him Fullerian Professor of Chemistry at the Royal Institution.

Although Faraday had little time for electromagnetism in the 1820s, he did dabble in it occasionally. Like other researchers of his day, he reasoned that if an electric current could produce magnetism, then magnetism ought to produce an electric current. But nobody could find any evidence for such an effect. In 1825, returning briefly to the puzzle, Faraday took two pieces of wire a metre or so long, tied them together side by side with only a single sheet of paper separating them, and connected one of the wires to a battery, so that a current flowed along it. An instrument set up to measure any current flowing in the other wire (a galvanometer, named in honour of Galvani) did not even flicker, and Faraday went back to his other work.

Some of that other work had an influence on how Faraday developed his ideas about the way electric and magnetic forces are transmitted from one object to another. Between 1828 and 1830, Faraday used several of the Friday Evening Discourses to describe, and demonstrate, the work of Charles Wheatstone on sound and musical instruments. Wheatstone was notoriously shy and a reluctant lecturer, and Faraday enjoyed the opportunity to speak on his behalf.

One of the tricks they demonstrated was how the vibration of an object such as a thin metal plate could set a similar one some distance away vibrating in sympathy. It was a kind of acoustic induction, caused by the passage of sound waves as a vibration, through the air or the laboratory bench, from one plate to the other. A force was clearly being transmitted, not leaping instantaneously across space as a so-called 'action at a distance'. When Faraday returned to the study of electromagnetism in 1831, in his 40th year, with his work on glass completed, his innovatory lecture series at the RI established as a great success, and the RI's finances on a more secure footing than ever before, the idea of induction provided the key to his next great discovery.

There had been one important discovery about electromagnetic induction in the past ten years, but nobody had interpreted it correctly. In 1824, the Frenchman François

Arago had found that when a compass needle was suspended over a copper disc, and the disc was rotated, the needle was deflected. Peter Barlow and Samuel Christie independently noticed the same effect in England, using a rotating iron disc. But Arago's work made a greater impression on scientists at the time, because copper, unlike iron, is not magnetic. So how could a rotating copper disc generate a magnetic force?

Nobody realized why the compass needle was affected by the rotating disc. It was thought that it was the act of rotation that had somehow made the disc magnetic, independently of the presence of the magnetic compass. Even when it was found that rotating a magnet near the disc would make the disc turn in response, not even the greatest scientific minds of the time guessed that what was happening was that an electric current was being set up in the disc, and that this current had a magnetic field of its own which interacted with the magnet or compass needle nearby. The electric current is produced by the relative motion of the disc and the magnet (even when the magnet is only a tiny compass needle). So if the disc turns and the magnetic compass needle is still, there is an effect; and if a magnet moves relative to the disc while the disc is still, there is also an effect. It was the flurry of interest in electromagnetic induction caused by Arago's discovery that led to Faraday's unsuccessful experiments which paired wires in 1825. (Arago's wheel also provided the subject for a Friday Evening Discourse, given by Faraday on 26 June 1827.)

When he returned to the puzzle in 1831, Faraday devised a new piece of equipment – the induction ring. By then, it was clear that if an electric current was passed through a wire wound in a coil (actually a helix), the coil of wire would act like a bar magnet, with a north pole at one end and a south pole at the other. If the coil of wire was wound around an unmagnetized iron rod, the rod would become a magnet when the current was switched on. Faraday wound two coils of wire on to the opposite sides of a ring of iron about 15 centimetres across; the iron making up the ring was itself about 2 centimetres thick. He reasoned that if an electric current was passed through one of the coils, it would create some sort of magnetic tension in the iron ring, with the iron acting to focus the effect on the opposite side of the ring, where the other coil was wound. He hoped that the effect would be to induce an electric current in the second coil, which was connected to a galvanometer.

Everything was ready on 29 August 1831. The second coil was connected to the galvanometer, and then the first coil was hooked up to the current. To Faraday's surprise, the needle of the galvanometer flickered just at the moment the connection was made, but fell back to its zero position when a steady current was flowing in the first coil. When the connection to the battery was disconnected, the galvanometer flickered again, in the opposite direction. Faraday had discovered that the magnetic force associated with the first coil could indeed induce an electric current in the second coil – but only when the magnetic force was changing, either building up or dying away. The effect even worked, although less strongly, using two coils alone, with no iron core.

So far, Faraday had only made electricity

using electricity, even if the intermediary was the magnetism generated by the electricity in the first coil. In September 1831, though, he found that moving a magnet in and out of a circuit caused a brief pulse of electricity to flow, and in October he made the key discovery that when a coil was wound round a hollow paper cylinder and connected to a galvanometer, electricity could be made to flow by pulling a bar magnet in or out of the hollow centre of the coil. It was this discovery that quickly led to the development of electric generators, or dynamos, in which a magnet is made to rotate past a coil (or the coil is made to rotate past a fixed magnet) to generate electricity.

Faraday himself did not develop commercially useful generators, but he did invent a little dynamo in which a disc of copper, like the disc used in Arago's experiments, span between the magnetic poles of a large horseshoe magnet. Using springy metal strips to make sliding connections to the rotating disc, one near the centre of the disc and the other on the rim near the magnet, he obtained a steady electric current from the rotating disc.

The practical importance of all this is so obvious that we do not need to dwell on it at length. The story has often been told of how the Prime Minister, Robert Peel, visited the RI soon after Faraday's discovery of the dynamo effect, and asked what use it was. According to legend, Faraday replied 'I know not, but I wager that one day your government will tax it.' The discoveries, announced in a paper read to the Royal Society on 24 November 1831, raised Faraday to the very highest rung on the scientific ladder, in both the public eye and among his peers.

Faraday's own investigations then developed in two different directions – one public and the other, for a long time, private. On the one hand, he sought ways to use electricity, now readily available from generators, in chemistry. (He also took the trouble to carry out a series of experiments proving that electricity from all the different sources known at the time, including electric fish, static electricity, voltaic piles and dynamos, was indeed a single phenomenon.)

He used electricity to break down various compounds into their component parts (*elec-*

trolysis), and in a paper published in 1834 he introduced many terms that have become standard in industry and are familiar from our schooldays: *electrolyte*, for a liquid through which an electric current is passed; *electrodes*, for the two connections where electricity enters and leaves the liquid; *anode*, for the positively charged electrode; *cathode*, for the negatively charged electrode; *ions*, for the electrically charged particles in the electrolyte. Electroplating became established commercially in the 1840s, and electrolysis became an important method for the industrial production of some chemicals. Chlorine, for example, is produced by passing an electric current through salty water (brine, a strong solution of sodium chloride). But all of this was stuff that anybody could do, once Faraday had pointed the way.

From 1831 onwards, Faraday's own deep interest was in the nature of the forces of electricity and magnetism, and how these forces were communicated across space. When a magnet is placed underneath a sheet of paper, and iron filings are sprinkled on to the paper as it is gently tapped, the filings form curved lines linking the poles of the magnet. These curved lines show the path that a tiny magnetic pole would follow if it were free to drift between the two poles of the magnet, and this led Faraday to develop the idea of lines of force linking the magnetic poles, and forming a field of force extending outward from the magnet.

The idea of a line of force is particularly useful in picturing what happens during the induction of an electric current by a magnet. If a conductor is stationary relative to the magnet, it is stationary relative to the lines of force, and no current flows. But if the conductor moves relative to the magnet, it is cutting through the lines of force, and it is this relative motion of the lines of force across the conductor that induces a current in the conductor.

So why is there a brief flicker of current in the secondary coil of the induction ring when the current in the first coil is switched on or off? Because switching on the current causes a magnetic field to build up, pushing out lines of force from the coil; while the lines are pushing out through and past the secondary coil, they cause a current to flow. But once the field is steady, there is no induced current. Exactly the same thing happens in reverse when the current in the primary coil is switched off, and the field collapses.

Faraday first used the term 'line of force' in a scientific paper as early as 1831, and by then he was also convinced that as well as not acting in straight lines, the magnetic force might not be transmitted instantaneously, but would take time to propagate through space – it would have to, after all, to fit this picture of current being induced by lines of force spreading out from, or collapsing back into, the primary coil in the induction ring.

The idea was so revolutionary that he hesitated to publish it. But he wanted to establish his scientific priority, so on 12 March 1832 he wrote a note which was placed in a sealed and dated envelope in a safe at the Royal Society. It was opened only after his death. Among other things, the note expressed Faraday's belief that:

> when a magnet acts upon a distant magnet or piece of iron, the influencing cause (which I may for the moment call magnet-

> ism) proceeds gradually from the magnetic bodies, and requires time for its transmission . . . I am inclined to compare the diffusion of magnetic forces from a magnetic pole, to the vibrations upon the surface of disturbed water, or those of air in the phenomena of sound: i.e. I am inclined to think the vibratory theory will apply to these phenomena, as it does to sound, and most probably to light.

It would be more than twelve years before Faraday went public with these ideas, a delay caused partly by his work on electrochemistry, partly by his reluctance to publish such outrageous claims, and partly by a serious bout of ill-health, brought on by overwork, at the end of the 1830s.

Faraday's achievements in the years 1831 to 1838 were at an astonishing level for any scientist, and highly unusual for a man in his forties – the greatest original ideas in science usually come from people in their twenties or early thirties. The strain had been particularly hard on Faraday, though. For a long time, perhaps since as early as 1816, he had suffered from such a bad memory that he had to make detailed notes for himself to remind him of progress with his work. It all became too much in 1839, and he suffered a classic nervous breakdown, brought on by over-work. In one of his notes, complaining about doctors who refused to listen properly to what he had to tell them, he wrote:

> When I say I am not able to bear much talking, it means really, and without any mistake, or equivocation, or oblique meaning, or implication, or subterfuge, or omission, that I am not able; being at present rather weak in the head, and able to work no more.

There had, with hindsight, been signs of what might happen. The first biography of Faraday, written by his successor at the RI, John Tyndall, and published in 1868, tells us that:

> Underneath his sweetness and gentleness was the heat of a volcano. He was a man of excitable and fiery nature; but through

> high self-discipline he had converted the fire into a central glow and motive power of life, instead of permitting it to waste itself in useless passion . . . he completely ruled his own spirit.

In other words, he was a self-control freak. Faraday was also obsessive about timekeeping. It is a tradition, inherited from his habits, that the Friday Evening discourses at the RI, which are otherwise relatively informal occasions, begin precisely at the stroke of 9 o'clock, when the door flies open and the speaker rushes to his place.

Things were not all sweetness and light during the hectic period of work in the 1830s. One of Faraday's nieces, Margery Ann Reid (who was the daughter of Sarah's sister Elizabeth, and later married James, the son of Michael's brother Robert) lived with the Faradays in the 1830s. She later noted that 'when [Faraday was] dull and dispirited, as he sometimes was to an extreme degree, my aunt used to carry him off to Brighton, or somewhere, for a few days, and they generally came back refreshed and invigorated.'

At the end of the 1830s, though, something more than a few days in Brighton was needed. Faraday spent much of his convalescence in Switzerland, attended by Sarah and her brother George, and did not really return to research until 1845 – in his 54th year. By then, he was well established as the grand old man of British science. Perhaps acknowledging that he could not go on for ever, he had taken the opportunity of a Friday Evening Discourse in 1844, when he was almost fully recovered from his breakdown, to give his ideas about lines of force a public airing.

The subject of the lecture, on 19 January 1844, was really an attack on the atomic theory of matter. Faraday argued that there could be no real distinction between space and the hypothetical atoms, and proposed instead that these atoms were simply the centres of concentration of forces. Instead of thinking of an atom as something that was the source of a web of forces, actually creating those forces, Faraday asked his audience to accept that what was fundamentally real was the web of forces itself, and that the atoms were simply concentrations of the lines of force making up the web – knots in the field. This idea is remarkably similar to the picture in modern quantum field theory, in which only fields have independent existence, and all particles are products of the field. But it made very little impact at the time, in spite of Faraday's use of graphic imagery to make his case.

In a classic example of a 'thought experiment', Faraday made it clear that he was talking about all the forces of nature, not just electricity and magnetism. He asked people to imagine the Sun sitting alone in space. What would happen if the Earth was suddenly, by magic, placed in its position at the appropriate distance from the Sun? How would it 'know' that the Sun was there? Faraday argued that even before the Earth was put in its place, the Sun's influence would extend through space, and through that place, in the form of lines of force. When the Earth was dropped into the web of lines of force, it would respond instantly to the presence of the lines of force – to the gravitational field – at the location of the Earth itself. As far as the Earth is concerned, what matters is the nature of the field at the Earth's location, not the nature of the source of the field – if, indeed, the Sun could really be regarded as the source of the field. To Faraday, the field was the reality, and matter (even matter on the scale of the Sun) merely associated with places where the field was concentrated.

In 1846, Faraday returned to the theme in another Friday Evening Discourse. This time, he was standing in for a speaker who had failed to appear. Legend has it that Charles Wheatstone was supposed to speak about some of his work, on 10 April 1846, but had an attack of stage fright and ran off at the last minute, leaving Faraday to hold the fort. One of us is among the many writers that have been guilty of repeating this delightful tale without checking it out (*Schrödinger's Kittens*, Phoenix). Alas for the legend, according to the RI's records the speaker booked for that evening was James Napier, and he had sent his apologies a full week before the meeting. Although Faraday was a stand-in lecturer on that occasion, and did indeed spend much of the talk describing Wheatstone's work, he did know in reasonable time what he was letting himself in for.

Whatever the background, though, the audience could hardly have expected that Faraday would use the spare time at the end of the lecture to air his 'Thoughts on Ray-vibrations'. Now, he suggested that light could be explained in terms of the vibrations of the electric lines of force. In the published version of the lecture, he said:

> The view which I am so bold as to put forth considers, therefore, radiation as a high species of vibration in the lines of force which are known to connect particles, and also masses of matter, together. It endeavours to dismiss the aether, but not the vibrations.

The 'aether' was a hypothetical substance, thought to fill all of 'empty space' and to provide a medium through which light waves could move, like ripples moving across a pond. Faraday then spelled out the nature of the kind of wave he was talking about:

> [It] is not the same as . . . the waves of sound in gases or liquids, for the vibrations in these cases are direct, or to and from the centre of action, whereas the former are lateral.

Faraday pointed out that the propagation of light takes time, and that this fits the idea of a ripple moving along a line of force. He speculated that gravity must propagate in a similar way, and also take time to travel from one object to another.

It was this package of great ideas that Faraday passed on to the next generation of scientists. James Clerk Maxwell was to develop a complete wave theory of light in terms of electromagnetic vibrations, building on Faraday's speculations, in the 1860s. Maxwell's work put those speculations on a secure mathematical footing, and also led him to predict the existence of similar electromagnetic waves with longer wavelengths than light. These waves (now known as radio waves) were duly discovered by Heinrich Hertz, in 1888.

Even in his late fifties, Faraday continued

to carry out scientific research, in short bursts between renewed bouts of ill-health. He studied the magnetic properties of glass and flames, liquids, solids and gases, and he investigated the effect of magnetic fields on light. All this was good stuff, and much of it helped him to develop his ideas about fields and lines of force. But, like his earlier work in chemistry, any competent scientist could have done it, and somebody else surely would have done it if Faraday had not. The great crowning achievement of Faraday's career, although this was scarcely appreciated in his time (even in Tyndall's biography of Faraday the ideas about lines of force are relegated to a section headed 'Speculations'), was laying the foundations of field theory. In any modern description, his later scientific work inevitably appears anticlimactic.

In a way, the end of Faraday's life was anticlimactic. He never fully recovered from his illness at the end of the 1830s, and although he summed up his great work on field theory in the mid-1840s, pointing the way for Maxwell, his memory became so bad that he could not read of the developments in science, and could not keep pace with other scientists breaking new ground. He could easily remember events that had happened long before, but had trouble remembering the everyday details of his life, including the names of his friends.

The illness seems to have been progressive. At first, after his recovery from the attack at the end of the 1830s, he was almost fully restored, but successive minor attacks left him with increasing difficulties. After the mid-1850s he became increasingly confused, and it seems that by the 1860s he was more often in the confused state than in his previously clear state of mind.

This deterioration is usually attributed to advancing years – in 1861, after all, Faraday was 70. Some people have suggested that the problems may have been made worse by chemical poisoning, as a result of all the toxic substances that he had handled in his early years at the RI. But nobody has ever been able to match his exact symptoms with any known form of long-term poisoning of this kind. Whatever the causes, Faraday himself was painfully aware of his declining powers.

In 1862, he wrote to an old friend and long-time correspondent:

> Again and again I tear up my letters, for I write nonsense. I cannot spell or write a line continuously. Whether I shall recover – this confusion – do not know. I will not write any more. My love to you.
>
> ever affectionately yours,
>
> M. Faraday

But we do not want to end our story on too gloomy a note. If Faraday's mental powers did decline, they did so from a great height; and even late in life he was able to make important contributions in his role as a public scientist.

Or perhaps we should say public *servant*, because that is how Faraday always saw himself. His religious beliefs would not allow him to hold public office, or any position of great authority, and also discouraged him from accepting public honours. He turned down the offer of a knighthood, and the Presidency of the Royal Society – twice, in 1848 and 1857. On the second occasion he commented to Tyndall that 'if I accepted the honour which the Royal Society desires to confer on me, I would not answer for the integrity of my intellect for a single year'. Significantly, Faraday was also among the early supporters of the British Association for the Advancement of Science, founded at the end of the 1850s. This was intended as an organization of professional scientists as equals, without the trappings of a gentlemen's club and the air of social prestige that were such an important part of the Royal Society. In 1864, Faraday was invited to become President of the Royal Institution, a largely honorary position recognizing his great contribution to putting the RI on a secure footing, but even this could not be considered. It was, he said, 'quite inconsistent with all my life and views'.

But that life and those views were consistent with public service, and Sandemanians were specifically expected to be loyal to the Crown. This has led to one delightful, but probably apocryphal, story. Faraday was invited to take lunch with Queen Victoria one Sunday early in 1844. The trouble was, the Sandemanians required their members to attend church every Sunday, without exception. On the other hand, they required their members to be loyal servants of the Queen. The story goes that Faraday chose loyalty to the Crown over his loyalty to the church, and, worse, when reprimanded for his action he was far from penitent, and defended his action. This lack of penitence, it is said, is what led to him losing the status of Elder, which he had held since 15 October 1840, and which was not restored to him until 21 October 1860. He was even excluded from the church from 31 March to 5 May 1844.

The point of the story is that, to Faraday's contemporaries, it clearly rang true that he would take such an action, and be so stubborn in defending his corner. Unfortunately, there is not a shred of evidence that the lunch with Queen Victoria took place. But there is evidence of a mysterious schism in the London Sandemanian church at this time, with no fewer than nineteen members (including Faraday) briefly excluded for some obscure reason. They cannot, surely, all have been excluded for defending Faraday's right to have lunch with the Queen!

The kind of stubbornness and adherence to principle that made people believe the anecdote had surfaced more publicly in 1835, when Faraday was offered a Civil List pension – an income from the government, independent of his income from the RI. In a recent change of government, the Whig Lord Melbourne had replaced the Tory Robert Peel as Prime Minister. Peel had intended to award Faraday a pension in recognition of his scientific work. Melbourne felt under some obligation at least to discuss the matter with Faraday, but opened their conversation by saying that he regarded 'the whole system of giving pensions to literary and scientific people as a piece of gross humbug'. Faraday said that in that case he certainly did not want to have a pension, and took his leave.

The story caused a flurry of excitement in the Tory press, with both Melbourne and Faraday initially sitting on their high horses and refusing to compromise. Eventually, Melbourne offered an apology. It was made clear that the pension was something Faraday had earned on merit, and he accepted it.

As his ability to carry out lengthy pieces of

scientific research declined, Faraday devoted more time, after the mid-1840s, to public service. In 1844, ninety-five miners were killed in an explosion at Haswell Colliery, in County Durham. The government (once again under Peel) sent Faraday and the geologist Charles Lyell to find out what had happened. Faraday proved an able forensic scientist, taking a keen interest in both the forensic and the scientific aspects of the work, and noting that:

> testimony is like an arrow shot from a long-bow; the force of it depends on the strength of the hand that draws it. Argument is like an arrow shot from a crossbow, which has equal force though shot by a child.

The investigation found that the explosion had probably been caused by human carelessness. The miners were equipped with Davy lamps (devised by Faraday's old mentor at the RI), which could not cause explosions, even in a build-up of flammable gas. Faraday and Lyell learned that the miners had found they could light a pipe from such a lamp, and that smoking had been going on in places where dangerous gases were likely to build up.

Faraday was also called in to help advise on how to protect the paintings in the National Gallery from the filthy air of London in the 1850s, and to advise on the proposed restoration of the Elgin Marbles, housed in the British Museum. He was a scientific adviser to Trinity House, the organization responsible for lighthouses around the coast of Britain, from 1836 to 1865. For most of that time there is no record of his activities in this capacity, because the papers covering the first twenty years or so of the period were destroyed in an air raid on London in the Second World War. But we do know that he investigated the use of limelight and electric arc lights to replace the old oil-burning lamps. And he carried out a great deal of work in the field. In February 1860, in his 69th year, he reached a lighthouse in Kent only by scrambling across snow-filled fields, over hedges and walls. But as he happily reported to Trinity House, 'I succeeded in getting there and making the necessary inquiries and observations.'

In one of his last public duties, in 1862, Faraday gave evidence to a Royal Commission set up to investigate the education provided in the public schools, at that time still based almost entirely on Latin and Greek Classics. Faraday advised the Commission that not only was it desirable to teach science at this level, but it was possible to do so in a structured way, with the effectiveness of a pupil's mastery of the subject being tested by examinations.

This must have been a poignant occasion, because Faraday's greatest public achievements had been as an educator and teacher, in the broadest sense of the term. But by the time he gave that evidence, even those abilities had generally failed him. In 1861, he had conceded that he could no longer fulfil his obligations at the RI, and that, in particular, he could no longer deliver the lectures for young people that had been one of his great innovations. In October that year, just after his 70th birthday, he offered his resignation. The RI accepted that he could no longer give lectures, but asked him to stay on as Superintendent of the house and the laboratories. He gave his last Friday Evening Discourse on 20 June 1862, and severed his last connections with the RI in 1865.

By then, Michael and Sarah Faraday had already moved out of the building. In 1858, at the suggestion of Prince Albert, Queen Victoria had offered the couple a house at Hampton Court. The house was in a bad state of repair, and Faraday worried about the cost of making it properly habitable, but when the Queen learned of his concern, she paid for the renovations as well. But it was only from 1862 onwards that this became the Faradays' real home; before then, they still spent most of their time at the RI. Faraday increasingly lost his mental powers after 1865, sinking into senility. He died quietly, as he sat in his favourite armchair, on 25 August 1867. In accordance with his own wishes he was buried quietly in Highgate Cemetery. His headstone bears this simple inscription:

MICHAEL FARADAY
Born 22 September 1791
Died 25 August 1867

Afterword

We don't need to stress how important Faraday's work on electricity has been in shaping the modern world. But even some scientists are not as well aware as they should be of the way his field theory underpins the most fundamental aspects of theoretical physics today.

The basic concept of a field is easy enough to grasp. It grew from the idea of lines of force. You can think of lines of force as being like the ribs in a spider's web, reaching out from a centre – say, the location of a charged particle. Particles that interact with the lines of force are pulled into the centre along the lines.

But there are spaces in a spider's web. The field idea comes from the image of filling in those gaps to make a smooth surface. One widely used example of this idea is in Albert Einstein's general theory of relativity, which describes gravity in terms of fields. The gravity of the Sun, for example, is seen as making a dent in spacetime (as if the spider's web were a rubber sheet, poked in its centre).

Objects under the influence of the Sun's gravity are then pictured as marbles rolling in curved paths around the dent the Sun makes.

The big difference between modern field theory and Faraday's version, though, is that the modern theory is quantized. Faraday and Maxwell thought of the electromagnetic field as continuous, with light being conveyed by ripples in the field. But in the early part of the twentieth century it became clear that light could also be described in terms of little particles, field quanta, that came to be known as photons. Indeed, light *must* be described in this way if we are to explain many aspects of its behaviour. In this picture the classical idea of a field is replaced by the concept of particles, such as photons, carrying forces as they are exchanged between other particles, such as electrons.

But at the same time, things that we are used to thinking of as particles (such as electrons) can be thought of as waves. These waves can be described in terms of ripples in another kind of field (one field for each type of particle), and the particles themselves can be described as field quanta, just as photons can be described as the quanta of the electromagnetic field. In quantum field theory (the current bee's knees in physics, responsible for ideas such as quarks and quantum electrodynamics), there are *only* fields to worry about. All particles, whether matter particles like electrons or force carriers like the photon, are regarded as excited states of the appropriate fields.

Apart from the sometimes hairy mathematics involved in the modern calculations, this is exactly the situation described by Michael Faraday in his famous Friday Evening Discourse in 1844. There is no doubt that Faraday was not just a competent chemist, or merely a technician who was good with his hands and clever enough to invent the electric motor and the dynamo. He was a theorist of the first rank, in many ways the first modern physicist – not bad for a poor blacksmith's son and bookbinder's apprentice.

A brief history of science

All science is either physics or stamp collecting.

Ernest Rutherford

c. 2000 BC	First phase of construction at Stonehenge, an early observatory.
430 BC	Democritus teaches that everything is made of atoms.
c. 330 BC	Aristotle teaches that the Universe is made of concentric spheres, centred on the Earth.
300 BC	Euclid gathers together and writes down the mathematical knowledge of his time.
265 BC	Archimedes discovers his principle of buoyancy while having a bath.
c. 235 BC	Eratosthenes of Cyrene calculates the size of the Earth with commendable accuracy.

AD 79	Pliny the Elder dies while studying an eruption of Mount Vesuvius.
400	The term 'chemistry' is used for the first time, by scholars in Alexandria.
c. 1020	Alhazen, the greatest scientist of the so-called Dark Ages, explains the workings of lenses and parabolic mirrors.
1054	Chinese astronomers observe a supernova; the remnant is visible today as the Crab Nebula.
1490	Leonardo da Vinci studies the capillary action of liquids.
1543	In his book *De revolutionibus*, Nicholas Copernicus places the Sun, not the Earth, at the centre of the Solar System. Andreas Vesalius studies human anatomy in a scientific way.

c. 1550	The reflecting telescope, and later the refracting telescope, pioneered by Leonard Digges.
1572	Tycho Brahe observes a supernova.
1580	Prospero Alpini realizes that plants come in two sexes.
1596	Botanical knowledge is summarized in John Gerrard's *Herbal*.
1608	Hans Lippershey's invention of a refracting telescope is the first for which there is firm evidence.
1609–19	Johannes Kepler publishes his laws of planetary motion.
1610	Galileo Galilei observes the moons of Jupiter through a telescope.
1628	William Harvey publishes his discovery of the circulation of the blood.
1643	Mercury barometer invented by Evangelista Torricelli.

1656	Christiaan Huygens correctly identifies the rings of Saturn, and invents the pendulum clock.
1662	The law relating the pressure and volume of a gas discovered by Robert Boyle, and named after him.
1665	Robert Hooke describes living cells.
1668	A functional reflecting telescope is made by Isaac Newton, unaware of Digges's earlier work.
1673	Antony van Leeuwenhoeck reports his discoveries with the microscope to the Royal Society.
1675	Ole Roemer measures the speed of light by timing eclipses of the moons of Jupiter.
1683	Van Leeuwenhoeck observes bacteria.

1687	Publication of Newton's *Principia*, which includes his law of gravitation.
1705	Edmond Halley publishes his prediction of the return of the comet that now bears his name.
1737	Carl Linnaeus publishes his classification of plants.
1749	Georges Louis Leclerc, Comte de Buffon, defines a species in the modern sense.
1758	Halley's Comet returns, as predicted.
1760	John Michell explains earthquakes.
1772	Carl Scheele discovers oxygen; Joseph Priestley independently discovers it two years later.
1773	Pierre de Laplace begins his work on refining planetary orbits. When asked by Napoleon why there was no

	mention of God in his scheme, Laplace replied, 'I have no need of that hypothesis.'
1783	John Michell is the first person to suggest the existence of 'dark stars' – now known as black holes.
1789	Antoine Lavoisier publishes a table of thirty-one chemical elements.
1796	Edward Jenner carries out the first inoculation, against smallpox.
1798	Henry Cavendish determines the mass of the Earth.
1802	Thomas Young publishes his first paper on the wave theory of light. Jean-Baptiste Lamarck invents the term 'biology'.
1803	John Dalton proposes the atomic theory of matter.
1807	Humphry Davy discovers

	sodium and potassium, and goes on to find several other elements.
1811	Amedeo Avogadro proposes the law that gases contain equal numbers of molecules under the same conditions.
1816	Augustin Fresnel develops his version of the wave theory of light.
1826	First photograph from nature obtained by Nicéphore Niépce.
1828	Friedrich Wöhler synthesizes an organic compound (urea) from inorganic ingredients.
1830	Publication of the first volume of Charles Lyell's *Principles of Geology*.
1831	Michael Faraday and Joseph Henry discover electromagnetic induction. Charles Darwin sets sail on the *Beagle*.

1837	Louis Agassiz coins the term 'ice age' (*die Eiszeit*).
1842	Christian Doppler describes the effect that now bears his name.
1849	Hippolyte Fizeau measures the speed of light to within 5 per cent of the modern value.
1851	Jean Foucault uses his eponymous pendulum to demonstrate the rotation of the Earth.
1857	Publication of Darwin's *Origin of Species*. Coincidentally, Gregor Mendel begins his experiments with pea breeding.
1864	James Clerk Maxwell formulates equations describing all electric and magnetic phenomena, and shows that light is an electromagnetic wave.

1868	Jules Janssen and Norman Lockyer identify helium from its lines in the Sun's spectrum.
1871	Dmitri Mendeleyev predicts that 'new' elements will be found to fit the gaps in his periodic table.
1887	Experiment carried out by Albert Michelson and Edward Morley finds no evidence for the existence of an 'aether'.
1895	X-rays discovered by Wilhelm Röntgen. Sigmund Freud begins to develop psychoanalysis.
1896	Antoine Becquerel discovers radioactivity.
1897	Electron identified by J. J. Thomson.
1898	Marie and Pierre Curie discover radium.
1900	Max Planck explains how electromagnetic radiation is

	absorbed and emitted as quanta. Various biologists rediscover Medel's principles of genetics and heredity.
1903	First powered and controlled flight in an aircraft heavier than air, by Orville Wright.
1905	Einstein's special theory of relativity published.
1908	Hermann Minkowski shows that the special theory of relativity can be elegantly explained in geometrical terms if time is the fourth dimension.
1909	First use of the word 'gene', by Wilhelm Johannsen.
1912	Discovery of cosmic rays by Victor Hess. Alfred Wegener proposes the idea of continental drift, which led in the 1960s to the theory of plate tectonics.
1913	Discovery of the ozone layer

	by Charles Fabry.
1914	Ernest Rutherford discovers the proton, a name he coins in 1919.
1915	Einstein presents his general theory of relativity to the Prussian Academy of Sciences.
1916	Karl Schwarzschild shows that the general theory of relativity predicts the existence of what are now called black holes.
1919	Arthur Eddington and others observe the bending of starlight during a total eclipse of the Sun, and so confirm the accuracy of the general theory of relativity. Rutherford splits the atom.
1923	Louis de Broglie suggests that electrons can behave as waves.
1926	Enrico Fermi and Paul Dirac

	discover the statistical rules which govern the behaviour of quantum particles such as electrons.
1927	Werner Heisenberg develops the uncertainty principle.
1928	Alexander Fleming discovers penicillin.
1929	Edwin Hubble discovers that the Universe is expanding.
1930s	Linus Pauling explains chemistry in terms of quantum physics.
1932	Neutron discovered by James Chadwick.
1937	Grote Reber builds the first radio telescope.
1942	First controlled nuclear reaction achieved by Enrico Fermi and others.
1940s	George Gamow, Ralph Alpher and Robert Herman develop the Big Bang theory of the origin of the Universe.

1948	Richard Feynman extends quantum theory by developing quantum electrodynamics.
1951	Francis Crick and James Watson work out the helix structure of DNA, using X-ray results obtained by Rosalind Franklin.
1957	Fred Hoyle, together with William Fowler and Geoffrey and Margaret Burbidge, explains how elements are synthesized inside stars. The laser is devised by Gordon Gould. Launch of first artificial satellite, *Sputnik 1*.
1960	Jacques Monod and Francis Jacob identify messenger RNA.
1961	First part of the genetic code cracked by Marshall Nirenberg.

1963	Discovery of quasars by Maarten Schmidt.
1964	W.D. Hamilton explains altruism in terms of what is now called sociobiology.
1965	Arno Penzias and Robert Wilson discover the cosmic background radiation left over from the Big Bang.
1967	Discovery of the first pulsar by Jocelyn Bell.
1979	Alan Guth starts to develop the inflationary model of the very early Universe.
1988	Scientists at Caltech discover that there is nothing in the laws of physics that forbids time travel.
1995	Top quark identified.
1996	Tentative identification of evidence of primitive life in a meteorite believed to have originated on Mars.